Stateside Tours of Duty

Vickey Joins the Military Series

Book 5

Stateside Tours of Duty

Book Five

Table of Contents

First Stateside Tour of Duty

New Mexico

White Sands Missile Range (WSMR)

My first stateside duty followed my first European tour, leading me to a location that was a well-kept secret in the air defense world.

As I inquired about my upcoming assignment, I found that no one seemed to know anything about the station. Thus, I embarked on my journey without any knowledge of my new location.

Arriving at my new duty station as a Specialist promotable brought about a bit of an awkward welcome to my new unit. The imminent promotion to Sergeant added an extra layer of complexity, as did the fact that I was transitioning from a headquarters unit and had limited experience in my assigned military job. That posed a potential challenge, as upon promotion, I would be overseeing a squad of individuals who were already experienced in a role I had not yet fully grasped.

To preempt any potential issues, I wasted no time immersing myself in the job, determined to learn as much as possible before my promotion. I leaned on the expertise of my fellow Specialists and the Corporal in charge of the squad upon my arrival, listening attentively and absorbing as much knowledge as I could.

Upon settling into my new duty station at White Sands Missile Range (WSMR), I quickly understood why it was shrouded in secrecy. As the test site for new generations of air defense equipment, WSMR held a high-profile status. Being assigned to such an elite facility was a privilege afforded too only a select few.

My journey at WSMR commenced with the standard military in-processing procedures, familiarizing myself with the surroundings, personnel, and the intricacies of the job. It was imperative for me to become proficient in the tactical aspects of my role, especially considering the heightened scrutiny that accompanied the facility's high-profile status.

Several months after my arrival at WSMR, my promotion to Sergeant marked a significant milestone in my career. While it was an expected progression, the increased responsibility and expectations of a Sergeant were somewhat overwhelming. To fully embrace my transition into the Non-Commissioned Officer (NCO) corps, I actively sought out opportunities to develop my new role.

One of my first actions as a newly-promoted Sergeant was to become one of the originators of the Better Opportunities for Single Soldiers (BOSS) program at WSMR. Serving as one of the representatives for the White Sands BOSS program allowed me to contribute to the welfare and morale of my fellow soldiers, while also honing my leadership skills in a meaningful way.

The BOSS program at WSMR aimed to ensure that single soldiers received equitable benefits comparable to their married counterparts. Its success was monumental, leading to its implementation at numerous other duty locations across the military.

During my time at WSMR, I also experienced another first: temporary duty (TDY). TDY assignments became a regular occurrence at WSMR, given our role as test personnel. We were frequently required to leave our home station and embark on missions to remote locations to conduct live tests and operations.

The success of my first TDY mission was truly remarkable, earning accolades for the military and civilian crews involved, as well as the teams responsible for munitions. Our achievements were celebrated with the ultimate honor: our photo displayed on the hall of fame at the testing facility.

However, the euphoria of our triumph was swiftly overshadowed by a sobering reality. As our crew departed from the mission site, tragedy struck. While navigating down the hilly, unpaved road, I lost control of the vehicle, leading to a serious crash.

While thankfully no one was injured in the crash, I faced another significant first in the military: the responsibility of paying for lost or damaged government property. The financial repercussions amounted to a considerable sum, equivalent to half a month's pay for two months. That unexpected expense dealt a significant blow to my bank account.

Recovering mentally from the financial strain and the tarnish to my reputation took time. However, a few months later, I found solace and distraction in the joyous occasion of friends getting married. Their celebration provided a much-needed opportunity to move past the ordeal and focus on happier moments.

Some happiness after the gloom.

My tenure at WSMR concluded with a remarkable acknowledgment from the General for my volunteer efforts, underscoring the importance of community engagement within the military. Additionally, I was honored with a second promotion, achieving the rank of Staff Sergeant. Those milestones not only affirmed my dedication to service but also marked significant achievements in my military career.

Brigadier General and Mrs. Jerry L. Laws
request the pleasure of your company
at a Volunteer Recognition Coffee
for recognition of your volunteer efforts
with B.O.S.S.
on Tuesday, the twenty-fifth of April
from 10:30-11:30
214 Goddard
White Sands Missile Range, New Mexico
.S.V.P. By 17 April
78-7441

Second Stateside Tour of Duty

Texas, First Tour

Fort Bliss

Beginning my tour at my second stateside duty station after returning from Korea marked a significant transition in my military journey. While I had hoped to delay my assignment to the home of air defense, the needs of the Army dictated otherwise, leading me to Fort Bliss.

At Fort Bliss, my role underwent a substantial transformation. As a Staff Sergeant, I was entrusted with the responsibility of serving as an instructor. That new assignment tasked me with educating fresh recruits on the operation of one of the Army's premier weapon systems. Drawing upon my experiences from WSMR, I was well-equipped to excel in that instructional role, eager to share my knowledge and expertise with the next generation of soldiers.

Shortly after my arrival, I commenced training for my new role. However, my time in that position was unexpectedly brief, as I soon received orders from the Department of the Army (DA) directing me to report to Drill Sergeant School.

"What are they thinking," I thought.

"I have only been in the military six years, I'm not qualified to be a Drill Sergeant, the elite of the non-commissioned officer corps," I continued in thought.

A few days after receiving my orders, I contacted DA to confirm the directives and seek clarification regarding my selection for attendance at Drill Sergeant School. The explanation provided revealed that my military record demonstrated that I met all the qualifications for the position, and given the Army's shortage of qualified personnel, I was chosen to fill the role.

"But did you see my age in my record," I responded.

"Age is not a disqualifier," the counselor responded.

"Can I turn the assignment down; I really don't want to be a drill sergeant."

"You can say no to the assignment, but we will then assign you to Recruiting School. You will be attending one of these schools, you decide."

Given the circumstances, I weighed my options and chose what I perceived to be the lesser of two evils: attending Drill Sergeant School. At that juncture in my career, I wasn't mentally prepared to promote the Army to potential recruits. Thus, I made the decision to proceed with Drill Sergeant School.

During my tenure as an instructor, I frequently crossed paths with my former Warrant Officer from Korea. On each encounter, whether it was at the grocery store or while jogging around the track, he persistently urged me to submit my packet to become a Warrant Officer. He even offered to provide a letter of recommendation to support my application.

After several encounters with him and his persistent inquiries, I finally decided to submit a packet to Warrant Officer Candidate School. His unwavering encouragement and belief in my potential were influential factors in my decision. Although I wasn't confident about being selected, I respected his opinion and wanted to fulfill his expectations.

Before departing my role as an instructor to attend Drill Sergeant School, I managed to submit the packet. While I didn't anticipate being chosen, I felt it was worth the effort to have a positive response the next time he asked about it.

Fort Bliss
Drill Sergeant

Drill Sergeant School was akin to reliving Basic Training from scratch. As Drill Sergeant Candidates, our previous rank and experience counted for nothing; we endured the same grueling drills that we once went through and would later impose on new recruits.

For nine weeks, it felt like I was back in Basic Training, except this time I had to learn the intricate modules of each drill to effectively teach them to new recruits. The pressure to excel and avoid failure was immense.

Surviving the rigorous training was a personal triumph, but not everyone made it to graduation. Upon completion, I proudly donned the iconic Drill Sergeant hat, marking another significant milestone in my military career.

Drill Sergeant School graduate!

Experiencing the training at Drill Sergeant School was just the beginning; the real challenge came with the day-to-day responsibilities of being a Drill Sergeant. Interacting with trainees on a daily basis brought its own unique set of challenges, adding pressure not only to perform well individually but also to navigate my military career effectively.

Drill Sergeant on the trail!

Being a Drill Sergeant was an intense experience, with long hours, the responsibility of managing numerous trainees, and demanding physical expectations. It was a role that couldn't be fully understood until you walked in the Drill Sergeant's boots, which is why the duty was limited to two years.

During my second year as a Drill Sergeant, I received news that I had made the list for promotion to Sergeant First Class. That was yet another significant milestone in my military career, marking the transition to a senior non-commissioned officer role. The promotion brought with it increased responsibility, respect, and the potential for a long and fulfilling career in the Army.

The announcement of my promotion was a joyous surprise, and I was tasked with leading the unit in physical training to mark the occasion. I embraced the opportunity with enthusiasm, leading the run with spirited cadences.

However, months later, I received notification that I had been selected to attend Warrant Officer Candidate School (WOCS). That news brought a mixture of emotions, including excitement and apprehension, as it represented a new and challenging chapter in my military journey.

Reflecting on my time at Fort Bliss, I realized that it had been filled with unexpected twists and turns, each one bringing its own set of challenges and rewards. Becoming a Drill Sergeant was a daunting task that I initially approached reluctantly, but I ultimately took pride in achieving. Making the promotion list for Sergeant First Class was another significant accomplishment.

However, the prospect of attending Warrant Officer Candidate School (WOCS) brought about feelings of uncertainty. I had heard that the training was rigorous and demanding, and I wasn't sure if I was ready for another grueling experience, especially considering my age. Additionally, I questioned whether I possessed the necessary skills to excel in operating and maintaining one of the Army's most critical weapon systems.

Those doubts and uncertainties weighed heavily on my mind as I contemplated the next step in my military journey.

The looming prospect of attending Warrant Officer Candidate School (WOCS) filled me with apprehension and anxiety. The pressure not to fail the school was intense, and the thought of facing another pain-staking training experience gave me reason to pause. Despite the potential extension of my time at Fort Bliss if I were successful, I struggled to focus on anything beyond the immediate challenge of overcoming my fears and succeeding at WOCS.

Fort Bliss
Warrant Officer

As I embarked on my third stint in Basic Training, stepping into WOCS, I braced myself for the rigorous seven-week program ahead. Renowned as one of the Army's toughest schools, WOCS posed a new challenge, despite my prior experience at Drill Sergeant School.

Anticipating weeks of grueling demands, I mentally prepared myself for what lay ahead. Day one felt like a familiar scene from past

Basic Training experiences, marked by relentless yelling and daunting physical tasks—Deja vu indeed.

Throughout the course, we faced a barrage of physical and written assessments, alongside meticulous inspections. WOCS emphasized attention to detail, recognizing its paramount importance in molding technical expertise across various Army specialties. The scrutiny during inspections honed in on that critical aspect, leaving no room for oversight.

Another inspection!

While attention to detail took center stage, the physical demands of the training played a significant role in our transformation. Enduring grueling obstacle courses, enduring long runs, and facing what I dubbed "added" torture, reminiscent of Basic Training, were all part of the regimen.

Pull-ups were a requirement for graduation!

Demanding and grueling obstacle courses!

And what seemed like an impenetrable wall!

As we delved deeper into the training, I came to the realization that WOCS was a breeze compared to Drill Sergeant School. In terms of physical demands, WOCS ranked third among my Basic Training experiences. While it might have been a challenging course for those who hadn't endured Drill Sergeant School, it was relatively manageable for me.

Just a light day at physical training!

I pressed on through the weeks of training, eventually graduating and earning the rank of Warrant Officer (WO). As with my other Basic Training experiences, not everyone made it through to graduation. After completing the school, I returned to Fort Bliss to delve into the specifics of my new responsibility: operating and maintaining all aspects of one of the major air defense weapon systems.

Fort Bliss
First duty as a WO

Enduring months of training at Warrant Officer Basic Course (WOBC), I emerged as an expert on the air defense weapon system. Just before graduation, we received our assignments for our first duty station as Warrant Officers. For me, it was Fort Bliss. At that moment, what was supposed to be a three-year stint had transformed into the equivalent of two tours of duty.

Upon my arrival at my first duty station as Warrant Officer One (WO1), I was met with a noticeable lack of confidence in my technical expertise from my chain of command. It seemed as though my new Commander felt she had drawn the short straw in receiving a rookie WO for her first command. We were both assigned to a newly formed air defense unit, marking a significant leadership challenge for each of us—her first unit command and my debut as a maintenance officer. The pressure to succeed in our inaugural leadership roles was palpable.

My personal test felt particularly stressful and daunting because of my own doubts about my capabilities. New Warrant Officers are often referred to as "wobbly ones," a nickname that comes from our propensity to stumble as we navigate our new responsibilities. However, despite the initial struggles, we were expected to persevere, continuously recovering from setbacks until we mastered our tasks. That resilience was crucial for both me and my new Commander as we both aimed to prove ourselves in our new roles.

The initial two years as a Warrant Officer (WO) were considered the critical period for making and recovering from major mistakes. During that time, WOs were expected to absorb as much knowledge as possible about their specialized field. By the time they ascended to the rank of Chief Warrant Officer, they were expected to be experts—errors were less forgivable, and results were demanded, not excuses.

As an air defense technical expert for a weapon system crucial in a recent US conflict, the pressure on my role was immense. My new unit, still in its formative stages, underwent extensive training to ensure everyone was combat-ready and able to function effectively as a team.

The true test of our training and cohesion came after numerous field exercises. At the end of a 10-day field exercise, we conducted a live fire event. To heighten realism, we received orders to relocate from our field site to our launch site just hours before engaging targets. That scenario was designed to mimic real-life conditions closely, and we were kept unaware of the target types until the exercise began.

That entire experience was exhilarating. As a new unit, we demonstrated to our leadership that we were capable and ready to meet the challenges ahead. That not only tested our operational readiness but also our ability to adapt quickly and effectively under pressure.

Live fire success!

The exercise was a resounding success, and our unit earned the respect of the leadership. We concluded the exercise with traditional celebratory actions, which included dousing the leaders—a spirited way to mark the achievement and camaraderie developed through the intense training and successful execution.

The look of success!

After the successful live fire exercise, we prepared the equipment for our return home.

I concluded my first tour at Fort Bliss with a deployment to Saudi Arabia. Following my return to the states, I proceeded with a permanent change of station move, marking the next phase of my military career.

Third Stateside Tour of Duty

Texas, Second Tour

Fort Bliss

I returned to Fort Bliss for a second tour of duty, approximately four years after my initial stint there. Initially, I never anticipated that my early role as an air defender and a tester of new elements in the weapon system would evolve into becoming a maintainer and expert of the system.

Yet, years later, I found myself as the senior and most knowledgeable individual for the system at my new stateside unit. That return coincided with air defenders coming back from the Iraq War amid still-high tensions. Once again, I was part of a startup unit.

In that new capacity, as the senior Warrant Officer for an Air Defense Battalion, my responsibilities were comprehensive. I was tasked with ensuring that all missile crew members were trained and certified, ready to perform their duties effectively. My role extended beyond oversight; I served as an advisor and mentor to all Warrant

Officers within the battalion. Success for the individual units translated directly into success for the entire Battalion.

With the military still at a heightened alert status from the terrorist attack on US soil and the war in Iraq, it was imperative that air defense units remained ready to deploy. That necessitated that I stay sharp and well-informed about the latest threats to provide informed advice to the Commander on the best courses of action for training.

During that tour, I became deeply involved in training and field exercise planning. With September 11, 2001, still very much on everyone's mind, there would be many training exercises during that stateside tour of duty, but now I would be an essential part of training and field exercise planning.

My input was crucial regarding the capabilities of the individual units concerning their equipment status and personnel qualifications. My responsibilities now included coordinating and conducting evaluations and certifications, assisting in developing defense designs, and providing expert technical and tactical advice.

Reflecting on my journey, I marveled at how far I had come—from someone who once sought to change military roles to becoming a go-to expert in the air defense arena. It was a bit shocking, but I had become an expert, a go-to person for all things air defense.

My responsibilities had expanded significantly, encompassing the role of senior tactical director, where I controlled missile firings for subordinate units. The weight of the responsibilities was daunting, but I relied on my accumulated experience and knowledge to persevere.

As I would sit in planning meetings, the many details involved in executing a routine field exercise came into full focus. In the meetings, details of field sites, firing sites, re-supply operations, feeding, hygiene, communication, and more had to be worked out and I was a part of the planning. It was frightening, and many times overwhelming, but it was exhilarating. Every time the Battalion was tasked with a field exercise, whether training or actual, those details were always critical to the task, and I played a crucial role in ensuring that many of those elements were taken care of.

Returning to White Sands Missile Range (WSMR) for live fires brought back memories of past successes and prepared us for a major joint exercise later in the year.

During our final live fire exercise, on a brisk morning in the mountainous desert of New Mexico, the unit's crew members intently listened to the countdown for launching two missiles. Both missiles were destined skyward, aimed at intercepting a threat missile. Amidst the tense atmosphere, a sudden burst of white light illuminated against the backdrop of a clear blue sky marking the moment of launch. The missiles soared, accurately meeting their target, and successfully intercepted and destroyed the threat, demonstrating the precision and readiness of our unit.

That exercise was part of a series of live fire drills that offered invaluable practical learning experiences. Through those exercises, we honed the essential checks and validations needed for successfully tracking and engaging targets. Each successful mission bolstered the confidence and credibility of every team member, reinforcing their capability to execute their roles effectively.

The culmination of the rigorous training during that tour was a high-profile homeland defense mission. That critical test was designed to evaluate the air defenders' capability to protect against threats posed to the US on domestic soil. Such a mission not only demonstrated our preparedness but also underscored the strategic importance of our unit in national security. My experience was expanded when my unit was asked to participate in an experimental exercise that would test our weapon system's effectiveness in Homeland Security operations.

Participating in the exercise provided a unique opportunity to collaborate in a joint environment with Navy and Air Force personnel, significantly expanding my understanding of joint military operations. The primary goal was to enhance inter-service communication, ensuring that all forces could coordinate effectively. That setup tested the crews' abilities to assess and confirm the capabilities and likelihood of our Air Defense systems to protect the homeland from potential threats. Such collaborative exercises were crucial for refining our strategic approaches and operational readiness across different branches of the military.

The successful integration of all air defense systems within the unit with a Navy cruiser and Air Force fighters marked a historic achievement. That pioneering collaboration led to a complete mission success, and the strategies developed during that experimental exercise became templates for future similar exercises. That accomplishment not only demonstrated the effectiveness of inter-service cooperation but also enhanced our collective defensive capabilities.

My third stateside tour of duty was further marked by a deployment to Kuwait and a subsequent second tour of duty in South Korea. Those overseas assignments provided further opportunities to apply and refine the skills and strategies developed during my time in the US, expanding my experience in international settings and reinforcing the global scope of our defense responsibilities.

While in Kuwait, I was assigned as a tactical crew responsible for continuously monitoring the skies for potential threats. That role required vigilance and precision to ensure the security of the region. A few months after returning from Kuwait, I prepared for my next assignment—a second tour of duty in South Korea. During that tour, I took on the significant responsibility of serving as the unit liaison officer with the Air Force. That role was crucial for enhancing coordination and communication between the two branches, ensuring that operations were seamless and that strategic objectives were achieved effectively. That position not only deepened my understanding of joint operations but also highlighted the importance of inter-service cooperation in achieving broader defense goals.

Fourth Stateside Tour of Duty

Oklahoma

Fort Sill

During my fourth and final stateside tour of duty, I returned to the Air Defense School as an instructor, but with significantly higher rank and a new focus. That time, instead of teaching enlisted air defenders, I was tasked with instructing new Warrant Officers. The role was an opportunity to impart the knowledge and experience I had gathered over the years to the next generation of military leaders, shaping their approach to air defense responsibilities.

The location for that phase of my career was Fort Sill, which had become the new hub for Air Defense. The training school had been relocated there from Fort Bliss. That move coincided with my return from my second tour of duty in South Korea, marking a pivotal transition in both my career and the Air Defense training program. At Fort Sill, I was able to leverage my extensive field experience and tactical knowledge to enhance the training and readiness of new

Warrant Officers, ensuring they were well-prepared for the challenges ahead in their military careers.

My previous roles and the experiences I gained were crucial in securing my placement back at the training school. Notably, my involvement in the groundbreaking proof of concept mission for Homeland Security played a significant role in tilting the odds in my favor for a return to teaching. That mission showcased my capabilities in real-world applications of air defense strategies and demonstrated my ability to handle high-stakes situations effectively. Such experiences underscored my qualifications and readiness to instruct new Warrant Officers, ensuring that I could provide them with valuable insights and practical knowledge derived from frontline operations.

New home for Air Defense School, Fort Sill, OK.

However, with my return to the schoolhouse, my teaching focus shifted from traditional air defense skills to leveraging my writing background to enhance the communication abilities of new air defenders. I advocated for the inclusion of courses on communication skills, specifically writing and briefing, in the training curriculum for new officers. That initiative was inspired by my observations throughout my career, particularly while working with new Warrant Officers. I noticed a recurring gap in their ability to prepare written materials effectively, which highlighted a broader need for a refresher in writing skills.

Understanding the importance of clear and effective communication in military operations, I aimed to equip those officers with the necessary tools to articulate their ideas clearly and present information efficiently. That approach was not only about improving writing skills but also about enhancing their overall effectiveness in leadership roles, where communication was paramount.

As officers, air defenders were required to produce a significant amount of written correspondence, making it essential to enhance their writing skills early in their careers. Recognizing that necessity, my curriculum not only emphasized writing improvement but also aimed to sharpen the officers' verbal communication skills. Verbal abilities were critical, especially when it came to briefing leaders and articulating strategic concepts clearly and effectively.

By focusing on those areas, the training aimed to equip the new officers with comprehensive communication skills, enabling them to excel in both written and spoken aspects of their duties. That holistic approach to communication training was designed to bolster their confidence and competence, ensuring they could perform their responsibilities effectively and contribute positively to military operations.

Instructing the new Warrant Officers, I focused on the correct methods for presenting military briefings to leadership, a crucial skill for their roles. Returning to my roots in communication—both written and verbal—felt like a full-circle moment, rekindling my initial passions and expertise in the field.

Teaching writing to the Warrant Officers was particularly critical because effective communication was a fundamental aspect of their job. As noncommissioned officers transitioning into warrant roles, many may not have had continuous opportunities to refine their communication skills. That gap highlighted the importance of integrating robust communication training into their professional development.

By providing those officers with both foundational and advanced communication skills, I aimed to ensure they were not only effective in their operational duties but also capable of leading and influencing through clear and persuasive communication. That approach was designed to enhance their overall effectiveness and readiness as leaders in a demanding and dynamic military environment.

Training in writing and briefing skills was a means of equipping young officers for the crucial demands of their roles, offering them the tools necessary for success in their duties. It provided an avenue to ensure their effectiveness in their future positions.

My old Battalion moves to Fort Sill

The Air Defense School wasn't the sole significant air defense entity relocating to Fort Sill; my former battalion from Germany also underwent the transition. Those movements were integral components of the Base Realignment and Closure directive.

As additional air defense units and personnel arrived at Fort Sill, the school faced the task of modernizing its academic curriculum to align with the latest equipment advancements and operational tactics of the weapon system. Transitioning into the role of a curriculum developer became my primary responsibility. That new position not only enhanced my professional credentials but also revealed potential career pathways beyond my military service.

Juggling my military duties, teaching responsibilities, and curriculum development, I experienced a moment of clarity. Why not explore the possibility of transitioning into teaching adult learners after my military service? After all, that was the driving force behind my pursuit of a doctoral degree in education – to impart knowledge at a higher learning institution post-military. With that realization, I decided to take a leap and applied to teach writing, my area of expertise, at the nearby university.

Receiving a call back from the university offering me a position as an Adjunct Faculty to teach English Composition was beyond belief. I was overjoyed yet simultaneously gripped with fear. The prospect of entering the academic realm felt like stepping into uncharted territory – the civilian sector posed a whole new set of challenges for my abilities.

Armed with only the basic requirements for my class curriculum, I entered the classroom and embarked on teaching my inaugural session to a group of non-military students.

During my initial semester of teaching at the university, I realized that deficiencies in writing and speaking skills weren't exclusive to soldiers. That epiphany prompted me to transform my class from solely focusing on writing to a comprehensive communication course. Fortunately, my university supervisor had granted me significant autonomy over my curriculum, enabling me to integrate various communication components into the class.

The communication curricula I developed for both the military and university settings proved to be resounding successes. In the military, I started receiving requests to deliver communication skills training to other air defense students. Meanwhile, at the university, the increasing enrollment in my class served as a testament to its effectiveness. I found myself elevated to the status of a communication instructor rockstar, in high demand for all new teaching cycles, whether in the military or at the university.

For several years, I juggled teaching responsibilities with both the military and the university concurrently. However, as my fourth year at Fort Sill approached, I found myself confronted with two significant decisions: whether to prolong my military service and whether to continue instructing at the university. Before reaching a conclusion, I had to carefully contemplate various factors.

Prior to making any definitive decisions, I felt compelled to explore several possibilities. One significant opportunity that could greatly influence my next steps was the chance to assume the role of Chief of the Air Defense Warrant Officer Branch. I believed that my achievements as a Warrant Officer would position me favorably for consideration for that responsibility. That role would entail overseeing the movement and assignments of air defense Warrant Officers, offering a substantial leadership opportunity.

Aware that no woman had previously held the position, I acknowledged that securing the role of Chief of the Air Defense Warrant Officer Branch was a long shot. Nevertheless, feeling I had nothing to lose, I reached out to speak with the current Chief to gauge my prospects. His response, though not explicitly stating so, implied that I wasn't being seriously considered for the position. Moreover, with his impending retirement, it became apparent that I would likely be assigned to another overseas post, a prospect that didn't excite me, especially considering my advanced age.

After careful consideration, I resolved that retirement from the military was the most fitting course of action for me. I communicated my decision to my chain of command and initiated the paperwork to conclude my military career. Alongside that choice, I also opted to terminate my employment with the local university. I informed the Chief of the English department of my decision not to return for the next semester.

My final year at Fort Sill, I took part in a military retirement ceremony, marking the end of my active duty service. In the remaining time I had, I wrapped up the winter semester at the university before returning to Texas to embark on the next chapter of my life –

transitioning to civilian life after dedicating over 20 years to serving as a soldier.

Entering civilian life after the military posed a significant uncertainty, yet I approached that next endeavor with the same tenacity and perseverance that had propelled me through previous pivotal moments in my life.

Made in the USA
Las Vegas, NV
25 October 2026

56549165R00026